IFLA

International Federation of Library Associations and Institutions
Fédération Internationale des Associations de Bibliothécaires et des Bibliothèques
Internationaler Verband der bibliothekarischen Vereine und Institutionen
Международная Федерация Библиотечных Ассоциаций и Учреждений
Federación Internacional de Asociaciones de Bibliotecarios y Bibliotecas
国际图书馆协会与机构联合会
الاتحاد الدولي لجمعيات ومؤسسات المكتبات

About IFLA **www.ifla.org**

IFLA (The International Federation of Library Associations and Institutions) is the leading international body representing the interests of library and information services and their users. It is the global voice of the library and information profession.

IFLA provides information specialists throughout the world with a forum for exchanging ideas and promoting international cooperation, research, and development in all fields of library activity and information service. IFLA is one of the means through which libraries, information centres, and information professionals worldwide can formulate their goals, exert their influence as a group, protect their interests, and find solutions to global problems.

IFLA's aims, objectives, and professional programme can only be fulfilled with the cooperation and active involvement of its members and affiliates. Currently, approximately 1,600 associations, institutions and individuals, from widely divergent cultural backgrounds, are working together to further the goals of the Federation and to promote librarianship on a global level. Through its formal membership, IFLA directly or indirectly represents some 500,000 library and information professionals worldwide.

IFLA pursues its aims through a variety of channels, including the publication of a major journal, as well as guidelines, reports and monographs on a wide range of topics. IFLA organizes workshops and seminars around the world to enhance professional practice and increase awareness of the growing importance of libraries in the digital age. All this is done in collaboration with a number of other non-governmental organizations, funding bodies and international agencies such as UNESCO and WIPO. IFLANET, the Federation's website, is a prime source of information about IFLA, its policies and activities: www.ifla.org

Library and information professionals gather annually at the IFLA World Library and Information Congress, held in August each year in cities around the world.

IFLA was founded in Edinburgh, Scotland, in 1927 at an international conference of national library directors. IFLA was registered in the Netherlands in 1971. The Koninklijke Bibliotheek (Royal Library), the national library of the Netherlands, in The Hague, generously provides the facilities for our headquarters. Regional offices are located in Rio de Janeiro, Brazil; Pretoria, South Africa; and Singapore.

IFLA Series on Bibliographic Control Vol 34

Functional Requirements for Authority Data

A Conceptual Model

Edited by Glenn E. Patton

IFLA Working Group on
Functional Requirements and Numbering
of Authority Records (FRANAR)

Final Report
December 2008

Approved by the Standing Committees of the
IFLA Cataloguing Section and IFLA Classification
and Indexing Section
March 2009

K · G · Saur München 2009

IFLA Series on Bibliographic Control
edited by Sjoerd Koopman

The "IFLA Series on Bibliographic Control" continues
the former "UBCIM Publications – New Series".

Bibliographic information published by the Deutsche Nationalibliothek
The Deutsche Nationalbibliothek lists this publication in the Deutsche Nationalbibliografie;
detailed bibliographic data is available in the Internet at http://dnb.d-nb.de.

∞
Printed on acid-free paper / Gedruckt auf säurefreiem Papier

K. G. Saur Verlag, München
An Imprint of Walter de Gruyter GmbH & Co. KG

Printed in the Federal Republic of Germany by Strauss GmbH, Mörlenbach

ISBN 978-3-598-24282-3

Contents

Introduction

The Working Group on Functional Requirements and Numbering of Authority Records (FRANAR) was established in April 1999 by the IFLA Division of Bibliographic Control and the IFLA Universal Bibliographic Control and International MARC Programme (UBCIM). Following the end of the UBCIM Programme in 2003, the IFLA-CDNL Alliance for Bibliographic Standards (ICABS) took over joint responsibility for the FRANAR Working Group with the British Library as the responsible body.

The Working Group has three terms of reference:

1) To define functional requirements of authority records, continuing the work that the "Functional requirements of bibliographic records" for bibliographic systems initiated;

2) To study the feasibility of an International Standard Authority Data Number (ISADN), to define possible use and users, to determine for what types of authority records such an ISADN is necessary, to examine the possible structure of the number and the type of management that would be necessary;

3) To serve as the official IFLA liaison to and work with other interested groups concerning authority files: <in*d*ecs> (Interoperability of Data in E-Commerce Systems), ICA/CDS (International Council on Archives Committee on Descriptive Standards; later, International Council on Archives Committee on Best Practices and Professional Standards), ISO/TC46 for international numbering and descriptive standards, CERL (Consortium of European Research Libraries), etc.

This document fulfills the first of these terms of reference and represents one portion of the extension and expansion of the *Functional Requirements for Bibliographic Records* model that was envisioned by the IFLA Study Group on the Functional Requirements for Bibliographic Records. The second term of reference, dealing with numbering, was dealt with by the Working Group in a separate document. The third of the terms of reference represents an ongoing task that has produced many valuable contacts, which have enriched this document and other aspects of this group's work and have provided the opportunity for the group to review a number of other documents about authority data that have been produced during the time of the group's work.

Because the *Functional Requirements for Authority Data* is indeed an extension and expansion of the FRBR model, the Working Group has identified potential improvements to the FRBR model itself as well as to definitions, etc. These will be proposed to the IFLA FRBR Review Group for their consideration. In addition, the group has identified other IFLA publications relating to authority data that may require revisions. Those potential revisions will also be submitted to the relevant groups for consideration.

The IFLA FRBR Study Group noted the need for further analysis of the "entities that are the centre of focus for subject authorities, thesauri, and classification schemes, and of the relationships between those entities." While the Working Group has included some aspects of subject authorities in the authorities model, it has not undertaken the full analysis that the FRBR Study Group envisioned. Since the FRANAR Working Group began its work, a new working group, Working Group on Functional Requirements for Subject Authority Records (FRSAR),[1] has been charged with that task.

March 2009

Members of the IFLA Working Group on Functional Requirements and Numbering of Authority Records

Françoise Bourdon *(Chair, 1999-2002)*
Bibliothèque nationale de France

Christina Hengel-Dittrich
Deutsche Nationalbibliothek, Germany

Olga Lavrenova
Russian State Library

Andrew MacEwan
The British Library

Eeva Murtomaa
National Library of Finland

Glenn E. Patton *(Chair, 2002-2009)*
OCLC, USA

Henry Snyder
University of California, Riverside, USA

Barbara Tillett
Library of Congress, USA

Hartmut Walravens
International ISBN Agency, Germany

Mirna Willer
University of Zadar, Croatia

Secretariat

Marie-France Plassard *(1999-2003)*
IFLA UBCIM Programme

Consultant

Tom Delsey
Ottawa, Canada *(2001-2009)*

Functional Requirements for Authority Data

1. Purpose

In libraries, in museums, or in archives, a catalogue is a set of organized data describing the information, content managed by an institution. Authority data represents the controlled access points and other information that institutions use to collocate works by a specific person, family, or corporate body, or the various editions of a title. Controlled access points include authorized forms and variant forms of name assembled by cataloguers to identify an entity. For the purposes of this study, only name and title entities are addressed fully; however, subject terms within catalogues are among the other entities commonly subjected to authority control. Authority control, which means both the identification of entities represented by controlled access points and the ongoing management of them, is integral to the functioning of a catalogue. Authority control is beneficial to cataloguers able to identify and distinguish between the controlled access points within a catalogue. More importantly, authority control benefits end users, enabling them to search any controlled form of an author's name or of a title to retrieve bibliographic resources within catalogues.

The primary purpose of this conceptual model is to provide a framework for the analysis of functional requirements for the kind of authority data that is required to support authority control and for the international sharing of authority data. The model focuses on data, regardless of how it may be packaged (e.g., in authority records).

More specifically, the conceptual model has been designed to:

- provide a clearly defined, structured frame of reference for relating the data that are recorded by authority record creators to the needs of the users of that data;
- assist in an assessment of the potential for international sharing and use of authority data both within the library sector and beyond.

2. Scope

The functional scope of the study was intentionally limited to the library sector, but the study has been conducted with a view to comparative analysis with other sectors.

For the purposes of this model, the users of authority data include both the authority data creators who create and maintain authority data and end users who use authority information either through direct access to authority data or indirectly through the controlled access points and reference structures in library catalogues, national bibliographies, etc.

For the purposes of this study, authority data is defined as the aggregate of information about a person, family, corporate body, or work whose name is used as the basis for a controlled access point for bibliographic citations or for records in a library catalogue or bibliographic database.

Conventionally, authority data is structured in accordance with guidelines and specifications, such as those set out in IFLA's *Guidelines for Authority Records and References (GARR)*[2] and/or cataloguing rules. In current practice, the authority record normally contains the authorized access point for the entity as established by the cataloguing agency as the default form for displays in its catalogue, as well as access points for variant forms of name and authorized access points for related entities. The authority record will also normally include information identifying the rules under which the controlled access points were established, the sources consulted, the cataloguing agency responsible for establishing the controlled access point, etc.

For the purposes of this study, however, there are no *a priori* assumptions made about the physical structure of authority data, nor are there any assumptions made as to whether the data are stored in an authority file that is separate from the catalogue or bibliographic file *per se*, or fully integrated with it.

At a high level, the conceptual model encompasses authority data of all types. The entity relationship diagram (section 3.3) and the entity definitions (section 3.4) are intended to reflect authority data for persons, families, corporate bodies, and geographic entities, or for entities represented in catalogues by title authority data (of the type of titles of work or work/expression and collective titles of works or works/expressions) or creator-title authority data, for subject authority data (subject terms, thesauri terms, and classification indicia), and authority data for various types of names and identifiers (standard numbers, codes, etc.) associated to those entities. However, the detailed analysis of entity attributes (section 4) and relationships (section 5) focuses only on library authority data for name controlled access points associated with persons, families, and corporate bodies, and for creator-title and title controlled access points associated with works, expressions, manifestations, and items. In those sections, attributes and relationships associated exclusively with other types of authority data (such as subject authority data) are not included. Management information that may be maintained as part of authority data, such as version control information or treatment decisions about the management of the resource described by the data, is also not included.

3. Entity-Relationship Diagram and Definitions

3.1 Entity-Relationship Methodology

The methodology used to build this conceptual model is the same entity analysis technique used in the *Functional Requirements for Bibliographic Records (FRBR)*[3]. As described in section 2.3 of *FRBR*, the first step is the identification of the key objects that are of interest to users of information in a particular domain. Each of these key objects, or *entities*, then serves as the focal point for a cluster of data. A model developed using these techniques also depicts the relationships between one type of entity and another type of entity.

Once the high-level structure for the model has been laid out by identifying the entities and the relationships between those entities, the next step is to identify the primary characteristics or *attributes* of each entity. At a more detailed level, the model can also depict the relationships that may exist between instances of entities.

The entity-relationship model described in the following section has been used in this study to assess the relevance of each attribute and relationship to the set of user tasks defined later in this document.

Entity versus Attribute: In designing any conceptual model, a key decision involves whether to make something an attribute or a separate entity. The outcome of this decision depends on the future use that is to be made of the attribute or entity. The developers of the FRBR model saw significant advantages in declaring persons and corporate bodies to be separate entities that could then be related to the other entities laid out in that model. Cataloguers have traditionally thought of controlling the names for persons and corporate bodies through authority data. Declaring persons and corporate bodies as entities enables much more flexibility in the controlled naming and eliminates redundancies that would occur if they were modeled as attributes. The names for these entities can then be controlled in an authority record and linked to other authority records or to bibliographic records or holdings records as needed. The FRBR decision to treat these as entities rather than as attributes has been carried over into this conceptual model of authority data. In addition, in this model, declaring names and identifiers as entities allows more clarity in showing the relationships between the entities on which authority data are focused and the names and/or identifiers by which those entities are known and on which the controlled access points for those entities will be based.

3.2 Diagramming Conventions

A rectangle represents an entity (i.e., an object of interest to users of authority data).

A dotted-line rectangle surrounding a group of two or more entities indicates that a relationship represented by an arrow contiguous with the dotted line may apply to any and/or all of the entities represented within the rectangle.

A single-headed arrow on a line represents a relationship in which any given instance of the entity at the opposite end of the line may be associated with only one instance of the entity to which the arrow is pointing.

A double-headed arrow on a line represents a relationship in which any given instance of the entity at the opposite end of the line may be associated with one or more instances of the entity to which the arrow is pointing.

3.3 Entity-Relationship Diagram

The fundamental basis for the conceptual model is illustrated in **Figure 1.** The model can be simply described as follows: Entities in the bibliographic universe (such as those identified in the *Functional Requirements for Bibliographic Records*) are known by names and/or identifiers. In the cataloguing process (whether it happens in libraries, museums, or archives), those names and identifiers are used as the basis for constructing controlled access points.

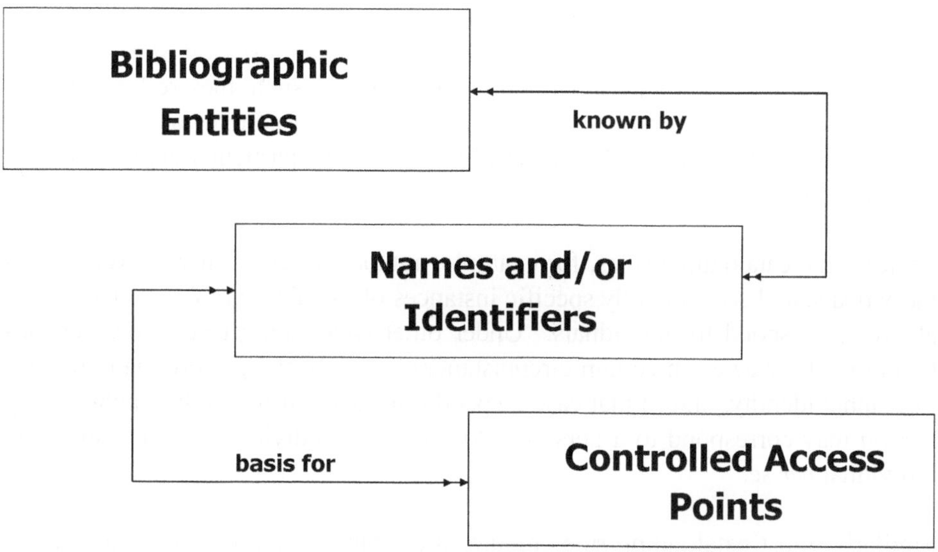

Figure 1: Fundamental Basis for the Conceptual Model

A fuller form of the conceptual model for authority data is presented graphically in the diagram in **Figure 2**. The entities depicted in the diagram are defined in section 3.4.

Entities: The entities depicted in the diagram represent the key objects of interest to the users of authority data. The resulting diagram represents a model of the relationships between those entities, not a model for an authority record, which is a specific application of authority data.

The entities on which authority data are focused (i.e., the ten entities defined in *Functional Requirements for Bibliographic Records – person, corporate body, work, expression, manifestation, item, concept, object, event,* and *place* – plus one additional entity – *family*) are depicted in the upper part of the diagram. The lower part of the diagram depicts the *names* by which those entities are known, the *identifiers* assigned to the entities, and the *controlled access points* based on those names and identifiers that are registered as authority data. The diagram also highlights two entities that are instrumental in determining the content and form of access points–*rules* and *agency*.

It is essential to note that the entities depicted in the upper part of the diagram (*person, family, corporate body, work, expression, manifestation, item, concept, object, event,* and *place*) are bibliographic entities. As such, they reflect intellectual constructs or concepts that are integral to the rules used to create library catalogues, and what is perceived as a specific instance of a particular entity type may vary from one set of rules to another.

Under some cataloguing rules, for example, authors are uniformly viewed as real individuals, and consequently specific instances of the bibliographic entity *person* always correspond to individuals. Under other cataloguing rules, however, authors may be viewed in certain circumstances as establishing more than one bibliographic identity, and in that case a specific instance of the bibliographic entity *person* may correspond to a persona adopted by an individual rather than to the individual *per se*.

Similarly, most cataloguing rules require the establishment of a new authorized form for a corporate body each time the body changes its name. In those cases, each specific instance of the bibliographic entity *corporate body* corresponds to the associated body only in relation to the period of time during which the body used that particular name. Consequently, the body as such (e.g., a legally incorporated enterprise, a government agency, a musical group) may be viewed as two or more instances of the bibliographic entity *corporate body*.

Rules for the construction of subject terms, on the other hand, may require the creation of a single authorized form to represent the body throughout the course of its existence. In that case, the specific instance of the bibliographic entity *corporate body* corresponds to the body *per se*.

Likewise, cataloguing rules may vary with respect to the treatment of *works*, *expressions*, *manifestations*, and *items*. The boundaries of a specific instance of the bibliographic entity *work* as defined by one set of cataloguing rules may differ from those defined by another set of rules. Hence, the two resulting instances of the bibliographic entity *work* may not be directly equivalent. Similarly, the application of one set of cataloguing rules may result in the recognition of a specific instance of the bibliographic entity *expression* that is not precisely the same as that recognized by the application of a different set of rules, etc.

Cataloguing rules and coding conventions may also differ in their treatment of fictional characters, supernatural or legendary beings, imaginary places, etc. In some cases, those entities will be viewed as *persons*, *families*, *places*, etc., but in other cases, they will be viewed as *concepts*. Similarly, certain events (ad hoc meetings, sporting events, etc.) may be viewed as *corporate bodies* under one set of cataloguing rules but simply as *events* under another set of rules.

Relationships between entities: The relationships depicted in the diagram **(Figure 2)** reflect the inherent associations between the various entity types. The lines and arrows connecting the entities in the upper part of the diagram with those in the lower part represent the relationships between *name* and *identifier* and the bibliographic entities with which they are associated (*person*, *family*, *corporate body*, *work*, *expression*, *manifestation*, *item*, *concept*, *object*, *event*, and *place*). A specific instance of any of those bibliographic entities may be known by one or more *names* ("has appellation" relationship), and conversely any *name* may be associated with one or more specific instance of any of the bibliographic entities ("is appellation of" relationship). Similarly, a specific instance of any one of the bibliographic entities may be assigned one or more *identifiers* ("is assigned" relationship), but an *identifier* may be assigned to only one specific instance of a bibliographic entity ("is assigned to" relationship).

It should be noted that the diagram **(Figure 2)** also depicts a relationship ("is associated with" relationship) that may apply between a *person*, a *family*, or a *corporate body*, on the one hand, and a *work*, *expression*, *manifestation*, or *item* on the other hand. The entity-relationship diagram in FRBR reflects the high-level relationships between entities *work*, *expression*, *manifestation*, and *item* (a *work* is "realized through" *expression*, etc.). It also depicts a number of specific relationships between *person* and *corporate body*, on the one hand, and *work*, *expression*, *manifestation*, and *item* on the other hand (a *work* may be "created by" a *person*, etc.). Those same relationships are conceptually valid for the FRBR entities that are represented in the conceptual model for authorities.

The relationships depicted in the lower part of the diagram **(Figure 2)** represent the associations between the entities *name* and *identifier* and the formal or structural entity *controlled access point*, and the association between that entity and the entities *rules* and *agency*. A specific *name* or *identifier* may be the basis for a *controlled access point* ("is basis for" relationship), and conversely a *controlled access point* may be based on a *name* or *identifier* ("is based on" relationship).

A *controlled access point* may be based on a combination of two *names* and/or *identifiers*, as in the case of a creator/title access point representing a *work* that combines the name of the author with the name (i.e., the title) of the work. The controlled access point will be centred on an instance of the *work* entity. However, in order to function effectively, the access point will need to reflect the relationship of that work not only to instances of the *name* entity (i.e., to the various titles by which the work is known) but its relationship to instances of the other entities depicted in the upper part of the diagram as well (most notably the person, family, or corporate body responsible for the work). The result is a creator/title access point.

Controlled access points may be governed by *rules* ("is governed by/govern" relationships), and those *rules* in turn may be applied by one or more *agencies* ("are applied by/applies" relationships). Likewise, *controlled access points* may be created by, or modified by one or more *agencies* ("is created/creates" and "is modified by/modifies" relationships).

These relationships are further described in section 5.2.

Relationships between instances of an entity: Other types of relationships that may operate between a specific instance of one bibliographic entity type and a specific instance of either the same or a different bibliographic entity type (e.g., a collaborative relationship between two specific persons or a membership relationship between a person and a corporate body) are not shown in the entity-relationship diagram. Relationships of that kind are discussed in section 5.3.

Other types of relationships that may operate between specific instances of the entities depicted in the upper part of the diagram (*person, family, corporate body, work, expression, manifestation, item, concept, object, event,* and *place*) and specific instances of the entity *name* are discussed in section 5.4.

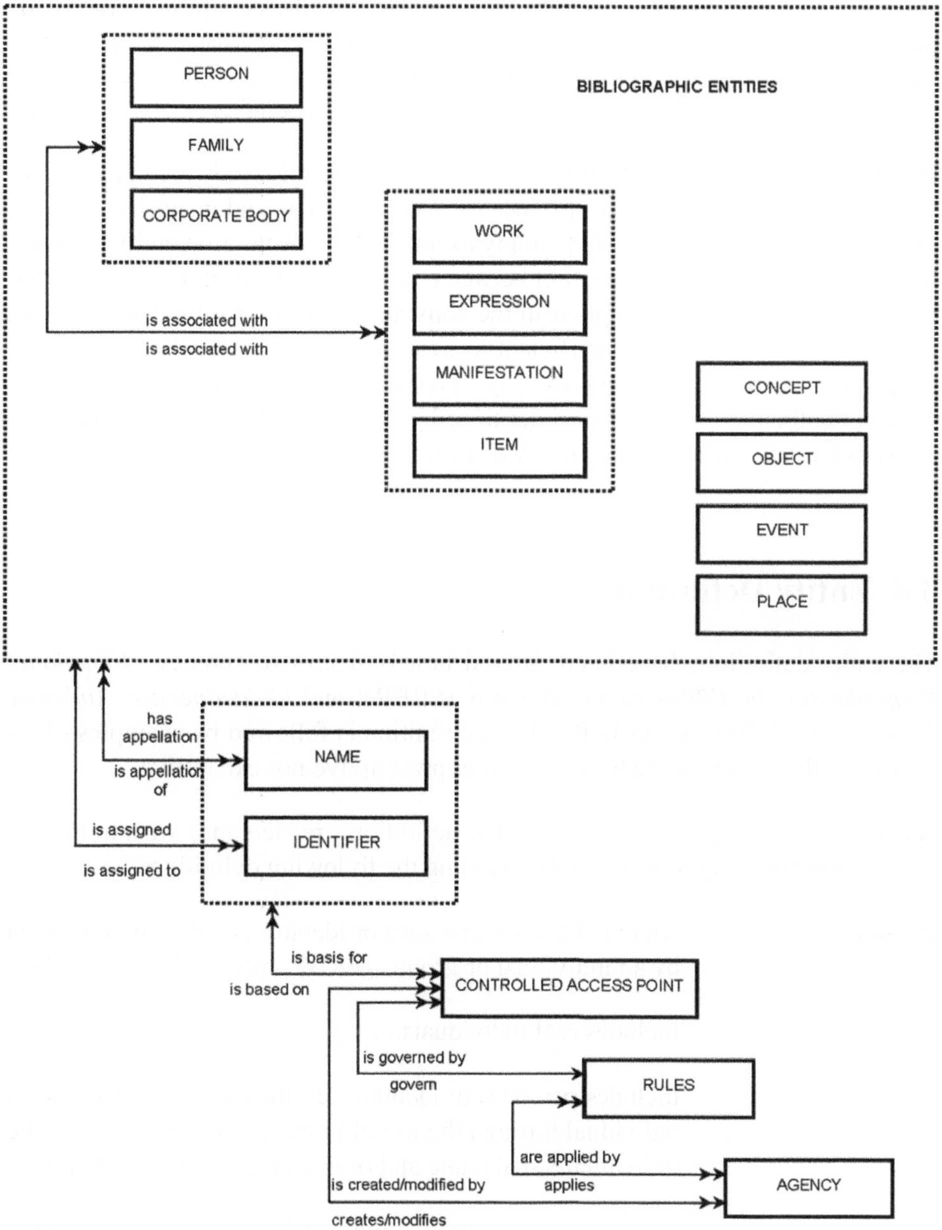

Figure 2: Conceptual Model for Authority Data

Other types of relationships that may operate between specific instances of the entity *controlled access point* (e.g., a parallel language or alternate script relationship between two specific access points) are not shown in the entity-relationship diagram. Relationships of that kind are discussed in section 5.5.

Relationships versus attributes: For the purposes of this study, certain types of association between a specific instance of one entity type and a specific instance of another entity type are treated simply as an attribute of the first entity. For example, in the case of an association between a *corporate body* and a *place* associated with the body that is implicit in the construction of a controlled access point for that body, it was deemed sufficient to treat the place associated with the body simply as an attribute of *corporate body*. That does not preclude the possibility of developing the model further in order to reflect that type of association more formally as a relationship between the two entities.

3.4 Entity Definitions

The entity definitions have been derived largely from two sources: *Functional Requirements for Bibliographic Records* (FRBR); and *Guidelines for Authority Records and References* (GARR). Each definition is followed by examples of instances of that entity; these lists are neither prescriptive nor exhaustive.

As an aid to relating the entities and their definitions, readers may wish to make a copy of **Figure 2** and refer to it when reading the following definitions.

Person

An individual or a persona or identity established or adopted by an individual or group. [FRBR, modified]

Includes real individuals.

Includes personas or identities established or adopted by an individual through the use of more than one name (e.g., the individual's real name and/or one or more pseudonyms).

Includes personas or identities established or adopted jointly by two or more individuals (e.g., Ellery Queen – joint pseudonym of Frederic Dannay and Manfred B. Lee).

Person (continued)	Includes literary figures, legendary figures, divinities, and named animals as literary figures, actors, and performers. Includes personas or identities established or adopted by a group (e.g., Betty Crocker). Includes appellations established by research (e.g., Bedford Meister, Meister E.S.) May include clusters of individuals who bear the same name, whenever it is not possible to establish a differentiated identity for each individual within the cluster. *Note: Cataloguing rules vary in their treatment of personas (i.e., bibliographic identities) established or adopted by the same individual. Some cataloguing rules recognize pseudonyms used by an individual as separate personas and an authorized form will be established for each persona with see also references connecting the resulting authorized forms of names for those personas. Other cataloguing rules recognize only one persona for an individual and the pseudonyms used by the person; one persona is the authorized form and the names of other personas are treated as variant forms of name for the individual.*
Family	Two or more persons related by birth, marriage, adoption, civil union, or similar legal status, or who otherwise present themselves as a family. Includes royal families, dynasties, houses of nobility, etc. Includes patriarchies and matriarchies. Includes groups of individuals sharing a common ancestral lineage. Includes family units (parents, children, grandchildren, etc.).

Family (continued)	Includes the successive holders of a title in a house of nobility, viewed collectively (e.g., Dukes of Norfolk).
Corporate Body	An organization or group of persons and/or organizations identified by a particular name acting as a unit. [FRBR, modified] Includes occasional groups and groups that are constituted as meetings, conferences, congresses, expeditions, exhibitions, festivals, fairs, etc. Includes musical performing groups, groups of visual artists, and dance companies producing collective work. Includes organizations that act as territorial authorities, exercising or claiming to exercise government functions over a certain territory, such as a federation, a state, a region, a local municipality. Includes organizations and groups that are defunct as well as those that continue to operate. Includes fictitious organizations or groups of persons. *Note: Cataloguing rules may differ with respect to the treatment of meetings, etc., depending on whether or not they are formally convened at a pre-determined date, whether or not they result in the issuance of a publication, etc. In some instances they may be treated as corporate bodies when they act as a group, but in other instances they may be treated simply as events (see also the entity definition for event).* *Note: Cataloguing rules generally treat distinctive identities of corporate bodies established through usage of different names over time as though they represented different but related corporate bodies.*

Work

A distinct intellectual or artistic creation (i.e., the intellectual or artistic content). [FRBR, as modified in the ICP Glossary[4]]

Includes works that are expressed in words, musical works, graphic works, photographic works, moving image works, cartographic works, three-dimensional works, data, computer programs, etc.

Includes legal works (laws, regulations, constitutions, treaties, court rules, court decisions, etc.), religious works (sacred scriptures, creeds, liturgical works, papal communications, etc.).

Includes collections and compilations of works, as well as individual works.

Includes component works (e.g., articles published in an issue of a periodical or the panels of an altarpiece).

Note: The concept of what constitutes a work and where the line of demarcation lies between one work and another may differ from one culture to another, from one discipline to another or from one cataloguing agency to another. Consequently, bibliographic conventions established by various cultures, national groups, or cataloguing agencies may differ with respect to the criteria they use for determining the boundaries between one work and another. The specifics of cataloguing rules may also have a bearing on what is recognized as a work.

Expression

The intellectual or artistic realization of a work in the form of alpha-numeric, musical, or choreographic notation, sound, image, object, movement, etc., or any combination of such forms. [FRBR]

Includes the specific words, sentences, paragraphs, etc., that result from the realization of a work in the form of a text (e.g., through translation).

Expression (continued)

Includes the particular notes, phrasing, etc., resulting from the realization of a musical work (e.g., through performance).

Includes the specific sequences of moving images, combined with sounds or not, that are available in a given version of a cinematic work.

Excludes aspects of physical form, such as typeface and page layout, unless they are integral to the intellectual or artistic realization of the work as such.

Note: The degree to which bibliographic distinctions are made between variant expressions of a work will depend to some extent on the nature of the work itself, and on the anticipated needs of users. Variations that would be evident only from a detailed analysis and comparison of expressions would normally be reflected in authority data only if the nature or stature of the work warranted such analysis, and only if it was anticipated that the distinction would be important to users.

Note: Controlled access points incorporating additions that reflect attributes of an expression (e.g., language of expression, form of expression) normally serve to collocate related expressions of the work as distinct from differentiating between individual expressions of the work.

Manifestation

The physical embodiment of an expression of a work. [FRBR, modified]

Includes manifestations in the form of books, periodicals, maps, posters, sound recordings, films, video recordings, DVDs, CD-ROMs, multimedia kits, PDF files, MP3 files, etc.

Note: If the physical embodiment of an expression of a work involves the production of only a single artifact (e.g.,

Manifestation (continued)

an author's manuscript or an original oil painting), the manifestation comprises only that single artifact. However, if the embodiment involves the production of multiple copies (e.g., in the publication of a book or the commercial production of a sound recording) the manifestation encompasses the entire set of physical copies produced. In the latter case, the entity described for cataloguing purposes is, in effect, an abstraction derived from characteristics of a single copy that are presumed to be shared by all copies in the set that comprises the manifestation.

Item

A single exemplar of a manifestation. [FRBR]

Includes items comprising a single physical object (one volume, cassette, etc.).

Includes items comprising more than one physical object (two or more volumes, compact discs, etc.) and electronic resources on the Web in multiple files.

Concept

An abstract notion or idea. [FRBR]

Includes fields of knowledge, disciplines, schools of thought (philosophies, religions, political ideologies, etc.), etc.

Includes theories, processes, techniques, practices, etc.

Note: The entity, Concept, *is defined as in FRBR without any extensions or clarification pending the work of the FRSAR Working Group.*

Note: For the purposes of this study, entities that are categorized as concepts are relevant only in the context of subject authority data.

Object

A material thing. [FRBR]

Includes animate and inanimate objects occurring in nature.

Object (continued)

Includes fixed, movable, and moving objects that are the product of human creation.

Includes objects that no longer exist.

Note: The entity, Object, *is defined as in FRBR without any extensions or clarification pending the work of the FRSAR Working Group.*

Note: For the purposes of this study, entities that are categorized as objects are relevant only in the context of subject authority data.

Event

An action or occurrence. [FRBR]

Includes historical events, epochs, periods of time.

Note: The entity, Event, *is defined as in FRBR without any extensions or clarification pending the work of the FRSAR Working Group.*

Note: Cataloguing rules may differ with respect to the treatment of meetings, etc. In some instances they may be treated simply as events, but in other instances they may be treated as corporate bodies (see also the entity definition for corporate body).

Note: For the purposes of this study, entities that are categorized as events but do not function as corporate bodies are relevant only in the context of subject authority data.

Place

A location. [FRBR]

Includes terrestrial and extra-terrestrial locations.

Includes historical as well as contemporary locations.

Includes geographic features.

Place (continued)

Includes geo-political jurisdictions

Note: The entity, Place, *is defined as in FRBR without any extensions or clarification pending the work of the FRSAR Working Group.*

Name

A character or group of words and/or characters by which an entity is known in the real world. [FRBR, modified]

Includes names by which persons, families, and corporate bodies are known.

Includes titles by which works, expressions, and manifestations are known.

Includes names and terms by which concepts, objects, events, and places are known.

Includes real names, pseudonyms, religious names, initials, and separate letters, numerals, or symbols.

Includes forenames (or given names), matronymics, patronymics, family names (or surnames), dynastic names, etc.

Includes given names of sovereigns, popes, etc., with associated roman numerals.

Includes names of families, clans, dynasties, houses of nobility, etc.

Includes names representing the successive holders of a title in houses of nobility, etc., viewed collectively (e.g., Dukes of Norfolk).

Includes names used by a corporate body at various periods in its history.

Includes names of associations, institutions, business firms, not-for-profit enterprises, etc.

Name
(continued)

Includes names of governments, government agencies, projects, programmes, government officials, delegations, legislative bodies, armed forces, etc.

Includes names of religious bodies, local churches, etc., religious councils, religious officials, provinces, dioceses, synods, etc., papal diplomatic missions, etc.

Includes names of conferences, congresses, meetings, etc.

Includes names of exhibitions, athletic contests, expeditions, fairs, festivals, etc.

Includes names of subordinate and related bodies.

Includes numbers that form an integral part of the name of a corporate body or event.

Includes trade names.

Includes titles of content, parts of content, compilations of content, etc.

Includes titles by which a work has become known.

Includes the title proper of the original manifestation of a work.

Includes titles by which a work is identified in reference sources.

Includes titles under which a work has been published.

Includes titles by which manifestations of an expression have become known.

Includes the title proper of the original manifestation of an expression.

Name (continued)

Includes titles by which manifestations of an expression are identified in reference sources.

Includes titles under which manifestations of an expression have been published.

Includes the title proper of the original edition of a manifestation.

Includes titles by which a manifestation is identified in reference sources.

Includes variant titles appearing on or in a manifestation.

Includes names and terms for events, objects, concepts, and places.

Note: Cataloguing rules may mandate the use of a place name to represent both the geographic territory per se and the government that has jurisdiction over that territory (e.g., the use of the name Massachusetts to represent both the territory covered by the state and the government of the state, the official name of which is The Commonwealth of Massachusetts.)

Note: Cataloguing rules may differ with respect to the elements they consider as integral parts of the name as distinct from elements they consider as additions to the name for the purposes of constructing controlled access points.

Identifier

A number, code, word, phrase, logo, device, etc., that is associated with an entity, and serves to differentiate that entity from other entities within the domain in which the identifier is assigned. An identifier can consist of an identifier string (i.e., a sequence of numeric and/or alphabetic characters assigned to an entity to serve as a unique identifier within the domain to which the identifier is assigned)

Identifier (continued)

and a prefix and/or suffix (a character or set of characters (numeric and/or alphabetic) appearing before or after an identifier string).

Includes identifiers such as social insurance numbers assigned by a government authority.

Includes personal identifiers assigned by other registration authorities.

Includes business registration numbers, registration numbers for charitable organizations, etc., assigned by a government authority.

Includes corporate body identifiers assigned by other registration authorities (e.g., ISBN publisher's prefix).

Includes standard identifiers assigned by registration authorities identifying content (e.g., ISRC, ISWC, ISAN).

Includes standard identifiers assigned by registration authorities identifying manifestations (e.g., ISBN, ISSN).

Includes thematic index numbers assigned to a musical work by the publisher, or a musicologist.

Includes catalogue raisonné numbers.

Includes identifiers for items assigned by repositories (e.g., shelf number).

Includes classification numbers referencing specific entities (e.g., a classification number assigned to a particular painting).

Includes registered trademarks.

Note: Some of these identifiers may be included in authority data but kept confidential for reasons of privacy.

Note: These identifiers are limited to those associated with entities and do not include record numbers assigned to authority records.

Controlled Access Point

A name, term, code, etc., under which a bibliographic or authority record or reference will be found. [GARR, modified]

Includes access points designated as authorized (or preferred) forms of names (i.e., authorized access points) as well as those designated as variant forms of name (i.e., variant access points).

Includes access points based on personal, family, and corporate names.

Includes access points based on titles (i.e., names) for works, expressions, manifestations, and items.

Includes access points consisting of a combination of two access points, as in the case of a creator/title access point for a work which consists of an authorized access point for the name of the creator combined with an authorized access point for the name (i.e., the title) of the work.

Includes access points based on names and terms for events, objects, concepts, and places.

Includes access points based on identifiers, such as standard numbers, classification indicia.

Note: Elements added to the name per se *(e.g., dates) form an integral part of the controlled access point.*

Rules

A set of instructions relating to the formulation and/or recording of controlled access points (authorized forms, variant forms or references, etc.).

Rules (continued)	Includes cataloguing rules and interpretations of those rules. Includes coding conventions.
Agency	An organization responsible for creating or modifying a controlled access point. The agency is responsible for application and interpretation of the rules it creates and/or uses. The agency may also be responsible for the creation and maintenance of identifiers within its domain. Includes libraries, national bibliographic agencies, bibliographic utilities, consortia, museums, archives, rights management organizations, etc.

4. Attributes

Attributes for each of the entities defined above are listed and defined in sections 4.1 through 4.16 below.

The attributes listed and defined below were derived from an examination of data identified and/or defined in FRBR, GARR, *UNIMARC Manual – Authorities Format*[5], *Mandatory Data Elements for Internationally Shared Resource Authority Records*[6] (MLAR), as well as in the *International Standard Archival Authority Record for Corporate Bodies, Persons and Families*[7] (ISAAR(CPF)). Thus, they may represent a broader list of possible attributes than may traditionally be recorded in library authority data. There is no intended implication that the attributes are mandatory or required for any application.

Other models, such as the *Encoded Archival Context*[8], have emerged more recently and have not been fully evaluated. This may be a topic for future investigation.

For the purposes of this model, attributes are defined at a "logical" level (i.e., as characteristics of the entities to which they pertain, not as specifically defined data elements). In certain cases, the characteristics of a given entity may change over time (e.g., the field of activity in which a person is engaged). The model does not make an explicit distinction between attributes that may change over time and those that are not subject to change. If attributes were defined at the data element level, they would be noted as being either repeatable or non-repeatable data elements, respectively.

Some of the attributes defined in the model could be considered as components of the entity rather than attributes of the entity. Considering them as components would be justified if it were necessary to define attributes for the components that differ from those of the aggregate entity or if the mapping of user tasks to the components differs from the mapping of user tasks to the aggregate entity. For purposes of this model, neither of these cases was deemed to apply.

In certain instances, the model treats an association between one entity and another simply as an attribute of the first entity. For example, the association between a *person* and the *place* in which the person was born could be expressed logically by defining a relationship ("born in") between *person* and *place*. However, for the purposes of this study, it was deemed sufficient to treat place of birth simply as an attribute of *person*. That does not preclude the possibility of developing the model further in order to reflect that association more formally as a relationship between the two entities.

If the model were extended to treat such associations as relationships between defined entities, it would be possible, in turn, to define attributes applying to the relationships themselves. For example, if the association between a *person* and a *place* in which the person resides were defined as a relationship ("resides in"), it would be possible then to define an attribute of the relationship to reflect the dates during which the relationship applied (i.e., the period during which the *person* resided in that *place*). Again, however, for the purposes of this study, it was deemed unnecessary to define that level of detail.

Note that because *name* and *identifier* have been defined in this model as entities in their own right, and have been linked through the "appellation" and "assigned" relationship, respectively, to the entities *person*, *family*, *corporate body*, *work*, *expression*, *manifestation*, *item*, *concept*, *object*, *event*, and *place*, the corresponding attributes (e.g., name of person, identifier for person) are not listed under the associated entity. For example, the FRBR attribute "Title of work" is, in this model, treated as an attribute of the Name by which that work is known. Through the relationship, however, the attributes of *name* and *identifier* function in the same way as if they were defined as attributes of the associated entity.

4.1 Attributes of a Person

Dates associated with the person	A date or dates associated with the person. [FRBR] Includes year or year/month/day of birth, death, etc. Includes year, period, or century of activity.
Title of person	Rank, office, nobility, honour, etc., associated with the person. [FRBR]

Title of person (continued)	Includes titles of rank or office (e.g., Major, Premier, Governor General). Includes titles of royalty and nobility, ecclesiastical titles, courtesy titles and titles of honour (e.g., Queen, Duke, Pope, Sir, Dame). Includes secular or religious information elements (e.g., Mrs., Swami, Professor).
Gender*	A gender by which the person is identified (e.g., male, female, unknown, other).
Place of birth*	The town, city, province, state, and/or country in which the person was born.
Place of death*	The town, city, province, state, and/or country in which the person died.
Country*	A country with which the person is identified.
Place of residence*	A country, state/province, city, etc., in which the person resides or has resided.
Affiliation*	A group with which the person is affiliated or has been affiliated through employment, membership, cultural identity, etc.
Address*	The current or former address of the person's place of residence, business, or employer. Includes street address, postal address, telephone number, e-mail address, etc. Includes the address of a World Wide Web site operated by the person, about the person or related to the person.

Language of person*	A language the person uses when creating an expression for publication, broadcasting, etc.
Field of activity*	A field of endeavour, area of expertise, etc., in which the person is engaged or was engaged.
Profession/occupation*	A profession or occupation in which the person works or worked.
Biography/history*	Information pertaining to the life or history of the person.
Other information associated with the person	Information elements other than title by which a person is known or identified. Includes elements, such as "Saint", "Spirit", etc. Includes elements, such as "Jr.", "III", etc., when associated with a person whose name includes a surname. Includes information associating the person with a particular work (e.g., "Author of Early Impressions"). Includes information associating the person with another person (e.g., "Follower of Rembrandt")

*Attributes marked with an asterisk represent additions to those identified in *Functional Requirements for Bibliographic Records.*

4.2 Attributes of a Family

Type of family	A categorization or generic descriptor for the type of family. Includes categorizations, such as clan, dynasty, family unit, patriarchy, matriarchy, etc.
Dates of family	Dates associated with the family.
Places associated with family	Information pertaining to places where the family resides or resided or had some connection.
Field of activity	A field of endeavour, area of expertise, etc., in which the family is engaged or was engaged.
History of family	Information pertaining to the history of the family.

4.3 Attributes of a Corporate Body

Place associated with the corporate body	A geographic place at any level associated with the corporate body. Includes countries, states, provinces, counties, cities, towns, regional municipalities, etc. Includes a location in which a meeting, conference, exhibition, fair, etc., is held. Includes location of headquarters.
Dates associated with the corporate body	A date or range of dates on which a meeting, conference, exhibition, fair, etc., is held, or a date with which the corporate body is otherwise associated. Includes dates of establishment of organizations, business firms, etc.

	Includes dates of existence of an organization, government, etc.
Language of the corporate body*	A language that the corporate body uses in its communications.
Address*	The current or former address of the corporate body's office(s). Includes addresses for the corporate body's headquarters and/or other offices. Includes street address, postal address, telephone number, e-mail address, etc. Includes the address of a World Wide Web site operated by the corporate body, about the corporate body or related to the corporate body.
Field of activity*	A field of business in which the corporate body is engaged, its area of competence, responsibility, jurisdiction, etc.
History*	Information pertaining to the history of the corporate body. Includes historical information pertaining to an organization, institution, etc.
Other information associated with the corporate body	Information as to incorporation or legal status of the body, or any term serving to differentiate the body from other corporate bodies, persons, etc. [FRBR] Includes types of incorporation, etc. (e.g., GmbH). Includes types of bodies or types of jurisdictions (e.g., Fraternal Order, County, Musical Group).

	Includes other characteristics distinctive to the corporate body (e.g., the name of the denomination of a local church).

*Attributes marked with an asterisk represent additions to those identified in *Functional Requirements for Bibliographic Records.*

4.4 Attributes of a Work

Form of work	A class to which the work belongs. [FRBR]
	Includes forms, genres, etc. (e.g., novel, play, poem, essay, biography, symphony, concerto, sonata, map, drawing, painting, photograph)
Date of the work	The first date (normally the year) associated with the work. [FRBR, modified]
	Includes single dates and ranges of dates.
	Includes date of creation.
	Includes date of first publication or release, in the absence of an ascertainable date of creation.
Medium of performance	The instrumental, vocal, and/or other medium of performance for which a musical work was originally conceived. [FRBR]
	Includes individual instruments, instrumental ensembles, orchestras, etc.
	Includes individual voices, vocal ensembles, choirs, etc.

Medium of performance (continued)	Includes other objects to be used in the performance of the work (e.g., spoons, washboard).
Subject of the work*	The subject aspects of the work and its content. Includes information about the subject of the work. Includes classification numbers.
Numeric designation	A number assigned to the work. [FRBR, modified] Includes serial numbers, opus numbers assigned by the composer of a musical work, etc.
Key	The set of pitch relationships that establishes a single pitch class as a tonal centre for a musical work. [FRBR] Includes the key in which the work was originally composed.
Place of origin of the work*	The country or other territorial jurisdiction or cultural area from which the work originated. Includes area of origin, etc.
History*	Information pertaining to the history of the work. Includes historical information pertaining to a work, including title changes for continuing resources.
Other distinguishing characteristic	Any characteristic that serves to differentiate the work from another work with the same title. [FRBR]

Other distinguishing characteristic (continued)	Includes parts of intellectual or artistic content. Includes musical incipits.

*Attributes marked with an asterisk represent additions to those identified in *Functional Requirements for Bibliographic Records.*

Note: The attributes of a work listed above include only those that are normally reflected in controlled access points or in other data elements recorded in authority records. They do not include other attributes of a work that may be reflected in bibliographic records, as identified in Functional Requirements for Bibliographic Records.

4.5 Attributes of an Expression

Form of expression	The means by which the work is realized. [FRBR] Includes mode or medium of expression (e.g., alpha-numeric notation, musical notation, movement notation, sound, image, or any combination thereof).
Date of expression	The date the expression was created. [FRBR] Includes single dates and ranges of dates. Includes date of writing, typing, inscribing a text, translation, etc. Includes date of performing a dramatic work, a musical work, etc.
Medium of performance of expression	The instrumental, vocal, and/or other medium of performance represented in the *expression* of a musical work. [FRBR] Includes individual instruments, instrumental ensembles, orchestras, etc.

Medium of performance of expression (continued)	Includes individual voices, vocal ensembles, choirs, etc. Includes other objects to be used in the performance of the work (e.g., spoons, washboard).
Language of expression	A language in which the work is expressed. [FRBR] Includes language(s) of the expression as a whole. Includes languages of individual components of the expression.
Technique	The method used to visualize a graphic image (e.g., engraving) or to realize motion in a projected image (e.g., animation, live action, computer generation, 3D) [FRBR] Includes statements of technique used to distinguish expressions of a work.
Other distinguishing characteristic	Any characteristic of the expression that serves to differentiate the expression from another expression of the same work. [FRBR] Includes names associated with variant versions of a work (e.g., King James version). Includes "edition" and version designations relating to the intellectual content of the expression (e.g., Version 2.1; abridged ed.).

Note: The attributes of an expression listed above include only those that are normally reflected in controlled access points or in other data elements recorded in authority records. They do not include other attributes of an expression that may be reflected in bibliographic records, as identified in Functional Requirements for Bibliographic Records.

4.6 Attributes of a Manifestation

Edition/issue designation	An information element appearing in the manifestation that normally indicates a difference either in content or in form between the manifestation and a related manifestation previously issued by the same publisher/distributor (e.g., second edition), or simultaneously issued by either the same publisher/distributor or another publisher/distributor (e.g., large print edition, British edition). [FRBR]
Place of publication/ distribution	A city, town, or other locality associated in the manifestation with the name of the publisher/ distributor. [FRBR]
Publisher/distributor	An individual, group, or organization named in the manifestation as being responsible for the publication, manufacturing, distribution, issuing, or release of the manifestation. [FRBR]
Date of publication/ distribution	The date (normally a year) of public release of the manifestation. In the absence of a date designated as the date of publication or release, a copyright date or a date of printing or manufacture may serve as a substitute. [FRBR]
Form of carrier	The specific class of material to which the physical carrier of the manifestation belongs (e.g., sound cassette, videodisc, microfilm cartridge). [FRBR]
Numbering	The designation of volume/issue, etc., and/or date appearing in a serial. [FRBR]

Note: The attributes of a manifestation listed above include only those that are normally reflected in controlled access points or in other data elements recorded in authority records. They do not include other attributes of a manifestation that may be reflected in bibliographic records, as identified in Functional Requirements for Bibliographic Records.

4.7 Attributes of an Item

Location of item*	The collection and/or institution in which the item is held, stored, or made available for access.
Custodial history of item*	The record of previous ownership of an item.
Immediate source of acquisition of item*	The source from which an item was directly acquired and the circumstances under which it was acquired.

*Attributes marked with an asterisk represent additions to those identified in *Functional Requirements for Bibliographic Records.*

Note: The attributes of an item listed above include only those that are normally reflected in controlled access points or in other data elements recorded in authority records. They do not include other attributes of an item that may be reflected in bibliographic records, as identified in Functional Requirements for Bibliographic Records, or in holdings records.

4.8 Attributes of a Concept

Note: No attributes are defined for the entity, Concept, pending the work of the FRSAR Working Group

4.9 Attributes of an Object

Note: No attributes are defined for the entity, Object, *pending the work of the FRSAR Working Group*

4.10 Attributes of an Event

Note: No attributes are defined for the entity, Event, *pending the work of the FRSAR Working Group*

4.11 Attributes of a Place

Note: No attributes are defined for the entity, Place, *pending the work of the FRSAR Working Group*

4.12 Attributes of a Name

Type of name

The category of a name.

Includes personal names, corporate names, names of families, trade names, and titles of works and manifestations.

Includes names of concepts, objects, events, and places.

Name string

A sequence of numeric and/or alphabetic characters or symbols that represents the name of an entity.

Scope of usage	The form of work associated with a particular name for a persona. Includes forms, genres, etc., (e.g., literary works, critical works, works on mathematics, detective novels) associated with a name used by an author.
Dates of usage	Dates associated with the use of a particular name established by a person, corporate body, or family.
Language of name	The language in which the name is expressed.
Script of name	The script in which the name is rendered.
Transliteration scheme of name	The scheme used to produce the transliterated form of the name.

4.13 Attributes of an Identifier

Type of identifier	A code or other designation indicating the type of identifier (i.e., the domain in which the identifier is assigned). Includes alphabetic strings identifying the numbering system (e.g., "ISBN", "ISSN", "ISRC") Includes symbols designating the type of identifier (e.g., ®)

4.14 Attributes of a Controlled Access Point

Type of controlled access point	A categorization or generic descriptor for the controlled access point. Includes terms and/or codes designating the type of access point (e.g., personal name access point, family name access point, corporate name access point, meeting name access point, territorial name access point, title access point, collective title access point or access points consisting of a combination of names, such as creator/title access points).
Status of controlled access point	An indication of the level of establishment of the authorized access point for the entity (e.g., provisional).
Designated usage of controlled access point	An indication of whether the rules under which the controlled access point was constructed designate it as an authorized (or preferred) form (i.e., authorized access point) or as a variant (or non-preferred) form (i.e., variant access points).
Undifferentiated access point	An indication that the controlled access point for the authorized form of name (i.e., authorized access point) is insufficiently precise to differentiate between two or more entities identified by the same name.
Language of base access point	The language in which the base access point is recorded.
Language of cataloguing	The language in which additions to the base access point are recorded.
Script of base access point	The form of graphic characters in which the data in the base access point are recorded.

Script of cataloguing

The form of graphic characters in which any additions to the base access point are recorded.

Transliteration scheme of base access point

An indication of the scheme used in transliterating the base access point.

Transliteration scheme of cataloguing

An indication of the scheme used in transliterating any additions to the base access point.

Source of controlled acess point

The publication or reference source used in establising the form of name or title on which the controlled access point is based.

Includes the title, publisher, date, etc., of the resource for which the controlled access point was originally created.

Includes the title, edition, etc., of the reference source(s) used to establish the conventional name or title.

Base access point

All elements in a controlled access point that are integral to the name or identifier that forms the basis for the access point.

Includes the name element in a controlled access point beginning with the name of a person, family, or corporate body.

Includes the phrase element in a controlled access point beginning with a phrase associated with a person that is used in lieu of a name as such when the name of the person is unknown (e.g., A Physician, Author of Early Impressions).

Includes the name element for a subordinate or related body in a controlled access point beginning with the name of a superior body.

Base access point (continued)

Includes the name element for a government agency, official, legislative body, court, etc., in a controlled access point beginning with the name for the territorial authority.

Includes the name element for a religious council, official, province, synod, etc., in a controlled access point beginning with the name for the religious body.

Includes the title element in a controlled access point for a work.

Includes a term designating form at the beginning of the title element of a controlled access point for a musical work or expression (e.g., Symphony, Concerto).

Includes conventional and collective titles at the beginning of the title element of a controlled access point for a work or expression (e.g., Treaties, Laws, Works, Selections, Piano music).

Includes the additional element(s) in a collective title controlled access point (i.e., a subsequent term or terms used to subdivide the collective uniform title)

Addition

A name, title, date, place, number, or other designation added to the base access point.

Sometimes referred to as a "qualifier."

Includes the fuller form of name added to the base access point.

Includes the title of nobility, title of royalty, or ecclesiastical title added to the base access point.

Addition
(continued)

Includes dates added to the base access point.

Includes the place name associated with a corporate body added to the base access point.

Includes a number associated with a corporate body or a musical work added to the base access point.

Includes a title of an adaptation or version of a work added to the base access point.

Includes the name and/or number of a section or part of a work added to the base access point.

Includes a term designating the form of a work added to the base access point.

Includes a term designating language of expression added to the base access point.

Includes a term designating the key in which a musical work was originally composed added to the base access point.

Includes a term designating the medium of performance for which a musical work was originally conceived added to the base access point.

Includes other designations associated with persons and corporate bodies added to the base access point.

Note: For the purposes of this study, terms indicating the role of the entity represented by the controlled access point in relation to a work, expression, manifestation or item described in a bibliographic record to which the access point is

Addition (continued)

appended (e.g., editor, translator, illustrator) are excluded from the scope of additions to controlled access points as they are recorded in authority data.

4.15 Attributes of Rules

Citation for rules

A citation for the cataloguing code, set of rules, instructions, etc.

Includes the title of the rules, etc., and, as necessary, the name of the body responsible for the rules, the edition designation, date, etc.

Note: The citation may also include a reference to a specific rule or rules within the cataloguing code, etc.

Rules identifier

An acronym or initials or alphanumeric code identifying the rules.

4.16 Attributes of an Agency

Name of agency

The name by which the agency is commonly known.

Includes the name of an agency responsible for creating the record (i.e., the originating source), an agency responsible for transcribing the record into machine-readable form, an agency responsible for modifying the record, or an agency responsible for issuing the record.

Includes acronyms or initials used in place of the agency's full name.

Agency identifier

An alphanumeric code identifying the agency.

Includes standard codes used as agency identifiers (e.g., ISO 15511, International Standard Identifier for Libraries and Related Organizations (ISIL) and MARC 21 Organization Codes).

Location of agency

The country, state/province, city, etc., in which the agency is located or headquartered, or the agency's address.

5. Relationships

5.1 Authority Relationships in the Context of the Model

The sections that follow describe four broad categories of relationships reflected in authority data. The first category (described in section 5.2) comprises the relationships that operate at a generic level between the entity types depicted in the entity-relationship diagram in section 3.3 (**Figure 2**).

The second and third categories (described in sections 5.3 and 5.4) comprise relationships commonly reflected in the reference structure of the authority record (i.e., in the references themselves, in information notes, instruction phrases, or cataloguer's notes). In some implementations, these relationships may be made by using coded data to generate display constants. The first of those categories comprises relationships between specific instances of the entities *person*, *family*, *corporate body*, and *work*. The second comprises relationships between a specific instance of the entity *person*, *family*, *corporate body*, or *work*, on the one hand, and a specific *name* by which the entity is known.

The fourth category (described in section 5.5) comprises the relationships between specific instances of the entity *controlled access point* reflected in the linking structures embedded in authority records (e.g., the links between fields carrying the same data recorded in different scripts).

5.2 Relationships Depicted in the High-Level Diagrams

The entity-relationship diagram in section 3.3 (**Figure 2**) shows the relationships that operate between the entities defined in the model at the highest level of generalization.

The relationships depicted in **Figure 2** represent the associations between bibliographic entities and the names by which those entities are known and the identifiers assigned to those entities. Persons, families, corporate bodies, works, expressions, manifestations, items, concepts, objects, events, and places "have appellation of" (i.e., are "known by") the names that are associated with them. Similarly, those entities may be "assigned" identifiers.

Figure 2 also depicts a generic relationship between persons, families, and corporate bodies, on the one hand, and works, expressions, manifestations, and items, on the other. A person, for example, may be associated with a work (as the author of the work, as the addressee of correspondence contained in the work, as the defendant in a trial reported in the work, etc.). Similarly, a person or corporate body may be associated with the expression of a work (e.g., as the performer of the work), with a manifestation (e.g., as the producer of the manifestation) or with an item (e.g., as the owner of the item). The generic relationship ("associated with") may be reflected in name/title authority records through the combination of a name access point representing the person, family, or corporate body and a title access point representing an associated work, expression, manifestation, or item.

The relationships depicted in the lower part of **Figure 2** represent the associations between names and identifiers, on the one hand, and the controlled access points "based on" those names and identifiers, the rules that "govern" the form and structure of controlled access points, and the agencies that "apply" the rules in "creating and modifying" controlled access points.

5.3 Relationships between Persons, Families, Corporate Bodies, and Works

Within authority data, there are a number of relationship types that operate between specific instances of the entities depicted in the upper part of **Figure 2** (*person, family, corporate body, work, expression, manifestation, item, concept, object, event, and place*).

Relationships in this category include those that operate between specific instances of entities of the same type (e.g., the relationship between one person and another, the relationships between one work and another) as well as those that operate between a specific instance of one entity type and a specific instance of a different entity type (e.g., the relationships between a person and a corporate body). These relationships are usually expressed in authority data as the relationship between the authorized names (preferred form of name) given to each entity. These relationships may be made explicit by means of information notes, instruction phrases, cataloguer's notes, or coded data used to generate display constants, or they may simply be implied through the "see also" reference relationship between one authorized form of name for each entity and another (see 5.4 Relationships between the Various Names of Persons, Families, Corporate Bodies, and Works and 5.5 Relationships between Controlled Access Points).

Table 1 highlights sample relationship types that operate between specific instances of the entity *person*, *family*, *corporate body*, or *work*, or between an instance of one of those entity types and an instance of a different entity type, as commonly reflected in name, title, and name-title authority records.

The relationship types identified in the table are intended to be representative only and do not aim to provide an exhaustive taxonomy of relationships in this category that may be reflected in authority data. It is recognized that specific applications will select relationships important to the application, so there also is no intention to make these relationships mandatory or optional. (Note that relationships in this category specific to subject authority data are not included in the table.)

In addition to the types of relationships listed in **Table 1**, "role" relationships are all covered by the "is associated with" relationship shown in the upper part of **Figure 2** and described in section 3.3. Applications may choose to subdivide these associations (or "roles") depending on the need for more specificity. "Roles" include terms, such as editor, illustrator, translator, defendant, typographer, previous owner, that may reflect the specific nature of the relationships between a person, family, or corporate body and a work, expression, manifestation, or item described or referred to in a bibliographic record. Such terms may be identifying data in bibliographic or authority records, but currently are typically added only to bibliographic records.

Definitions and examples for each of the relationship types identified in **Table 1** are given below in sections 5.3.1-5.3.7.

Other communities may use different terminology. The examples also present suggestions of how the conceptual model might be applied in a particular application of authority data. Examples are not given for those relationships not usually expressed in authority records.

Table 1: Relationships between Persons, Families, Corporate Bodies, and Works

Entity Type	**Sample Relationship Types**
Person ⇔ Person	▪ pseudonymous relationship ▪ secular relationship ▪ religious relationship ▪ official relationship ▪ attributive relationship ▪ collaborative relationship ▪ sibling relationship ▪ parent/child relationship
Person ⇔ Family	▪ membership relationship
Person ⇔ Corporate Body	▪ membership relationship
Family ⇔ Family	▪ genealogical relationship
Family ⇔ Corporate Body	▪ founding relationship ▪ ownership relationship
Corporate Body ⇔ Corporate Body	▪ hierarchical relationship ▪ sequential relationship
Work ⇔ Work	▪ equivalence relationship ▪ derivative relationship ▪ descriptive relationship ▪ whole/part relationship ▪ accompanying relationship (part-to-part) ▪ sequential relationship ▪ shared characteristic relationship

5.3.1 Relationships between Persons

Pseudonymous relationship

The relationship between a "real" person (i.e., an individual) and a persona(ae) or identity adopted by that individual through the use of one or more pseudonyms.

Example

This is usually expressed in authority data through links between the authorized names (preferred form of name) given to each entity (see also references) and/or information notes:

The relationship between the person known as Eric Arthur Blair and the persona he used for his literary works, George Orwell.

The relationship between the person known as Barbara Mertz and the two other personas adopted by her: Barbara Michaels and Elizabeth Peters.

Secular relationship

The relationship between a person and an identity that person assumes in a secular capacity.

Example

The relationship between the pope known as "Pope John XXIII" and that pope's secular identity "Angelo Giuseppe Roncalli".

Religious relationship

The relationship between a person and an identity that person assumes in a religious capacity.

Example

The relationship between the person known as Thomas Merton and that person's name in religion, Father Louis.

Official relationship

The relationship between a person and an identity that person assumes in an official capacity.

Example

The relationship between the person known as John Fitzgerald Kennedy and that person's official name as President of the United States, for the period 1961-1963.

Attributive relationship

The relationship between one person and another person (either a "real" individual or a persona or identity) to whom one or more works by the former have been attributed, either erroneously or falsely.

Example

Person | Attributed Person

This is usually expressed in authority data through links between the authorized names given to each entity (see also references) and/or information notes:

The relationship between the person known as Marcus Junius Brutus and the person known as Pseudo-Brutus whose works have been erroneously attributed to Marcus Junius Brutus.

Collaborative relationship

The relationship between two or more persons (i.e., one or more individuals) who collaborate in intellectual or artistic endeavours.

Example

Person A Person B Persona

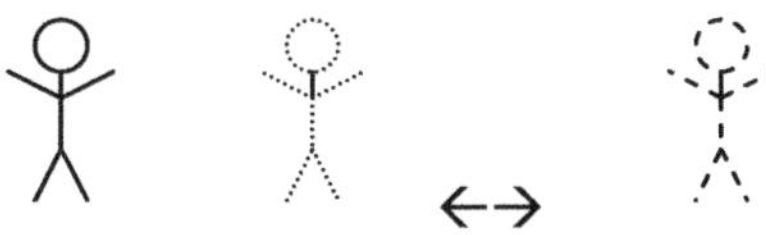

The relationship between the two persons known as Émile Erckmann and Alexandre Chatrain and their shared pseudonym Erckmann-Chatrain.

Sibling relationship

The relationship between two persons who have the same "parent(s)" (in an actual family or within a group). This type of relationship is often presented by virtue of the relationship to the parents or family or group to which the "siblings" belong. See "Membership relationship" in 5.3.2 Relationships between Persons and Families and 5.3.3 Relationships between Persons and Corporate Bodies.

Example

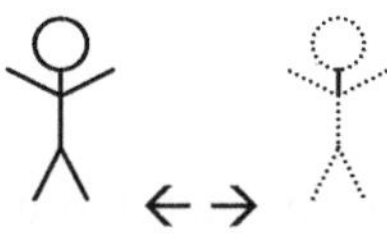

Siblings (have one or more parents in common)

This type of relationship is rarely expressed in bibliographic or authority data.

Parent/Child relationship

The relationship between two persons, one of whom is a "parent" or "child" of the other. This type of relationship is often presented by virtue of the relationship to the family or group to which the "child" belongs. See "Membership relationship" in 5.3.2 Relationships between Persons and Families and 5.3.3 Relationships between Persons and Corporate Bodies.

Example

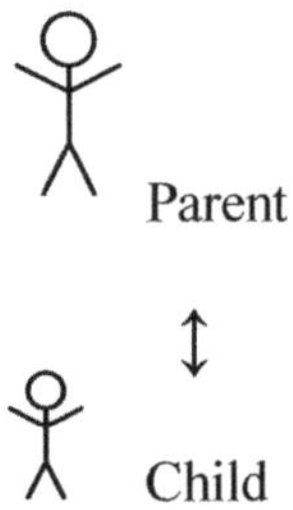

This type of relationship is rarely expressed in bibliographic or authority data.

5.3.2 Relationships between Persons and Families

Membership relationship

The relationship between a person and a family of which the person is a member.

Example

This is usually expressed in authority data through links between the authorized names given to each entity (see also references) and/or information notes:

> The relationship between Italian photographer known as Wanda Wulz and other members of her family known as the Wulz family.

5.3.3 Relationships between Persons and Corporate Bodies

Membership relationship

The relationship between a person and a corporate body in which the person is a member or with which the person is affiliated.

Example

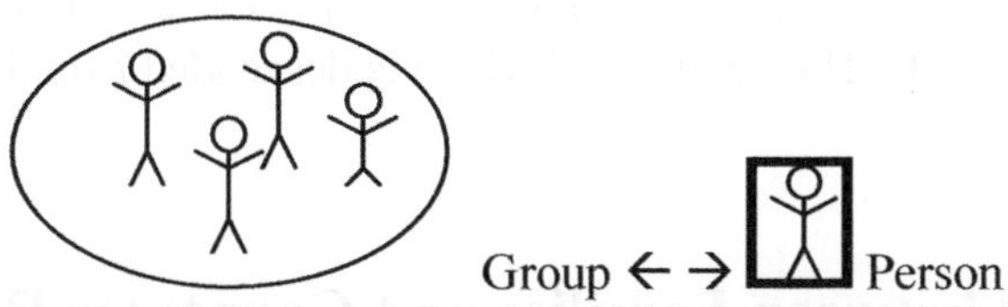

This is usually expressed in authority data through links between the authorized names given to each entity (see also references), cataloguer's notes, and/or information notes:

> The relationship between the three persons, Tomislav Alajbeg, Trpimir Alajbeg and Boris Budiša and the group known as Digitalis vulgaris.
>
> The relationship between the person known as Buddy Rich and the musical group known as the Buddy Rich Band.

5.3.4 Relationships between Families

Genealogical relationship

The relationship between one family and another that descended from the first or is the predecessor to the other.

Example

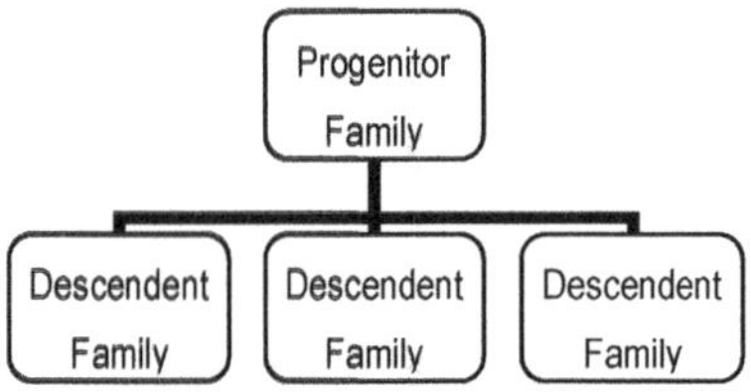

This is usually expressed in authority data through links between the authorized names given to each entity (see also references), cataloguer's notes, and/or information notes:

> The relationship between the family known as the House of Condé and the family known as the House of Conti which is descended from it.

5.3.5 Relationships between Families and Corporate Bodies

Founding relationship

The relationship between a family and a corporate body which the family created.

Example

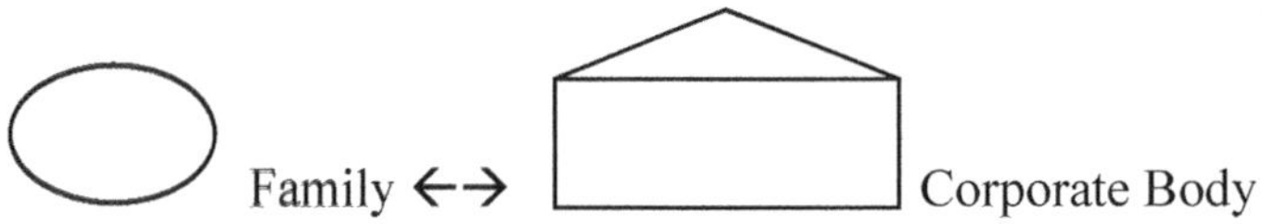

This is usually expressed in authority data through links between the authorized names given to each entity (see also references), cataloguer's notes, and/or information notes:

> The relationship between the family known as the Hall family and the corporation they created known as the Hall Family Foundation.

Ownership relationship

The relationship between a family and a corporate body which the family owns.

Example

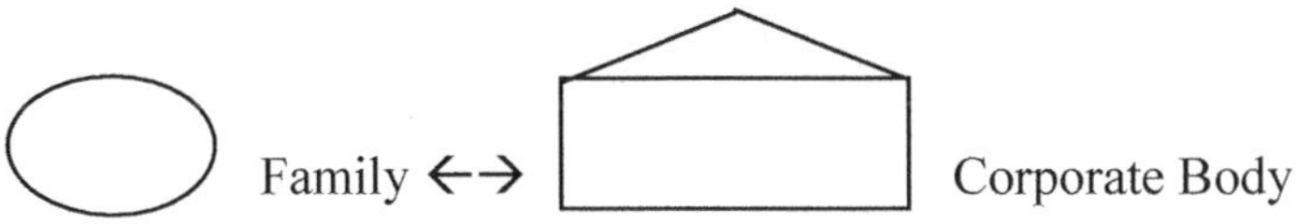

This is usually expressed in authority data through links between the authorized names given to each entity (see also references), cataloguer's notes, and/or information notes:

> The relationship between the family known as the Herlin family and the corporation they own known as Kone.

5.3.6 Relationships between Corporate Bodies

Hierarchical relationship

The relationship between two corporate bodies, one of which is subordinate to the other (within a corporate hierarchy, etc.)

Example

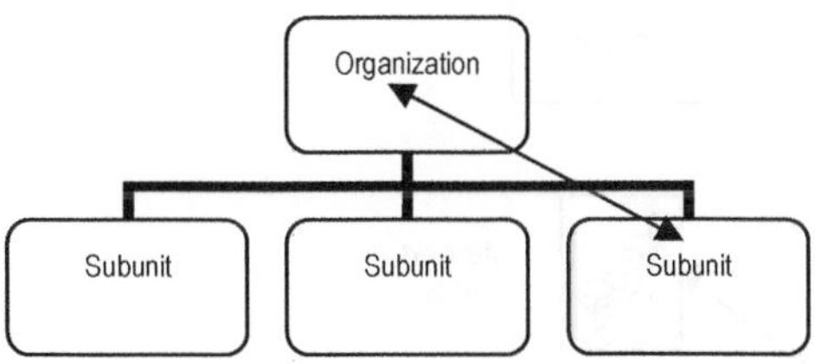

This relationship is usually expressed in authority data through the hierarchical structure of the authorized form of name given to the subordinate body (to include the preferred form of name of the higher level body), through see references, and/or explained through information notes or cataloguer's notes:

The relationship between the conference known as Journée de la kinésithérapie and the conference known as 11e Congrès de pneumologie de langue française, of which the Journée de la kinésithérapie was a part.

The relationship between the corporate body known as the Carl A. Kroch Library and corporate body known as Cornell University, of which the library is a part.

Sequential relationship

The relationship between two or more corporate bodies (i.e., organizations, meetings, conferences, etc.) that follow the first in order or time (e.g., predecessor/successor, splits, mergers, subsequent meetings/conferences, etc.).

Example

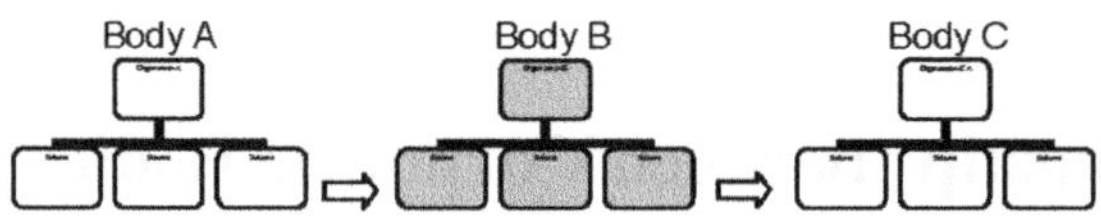

Or

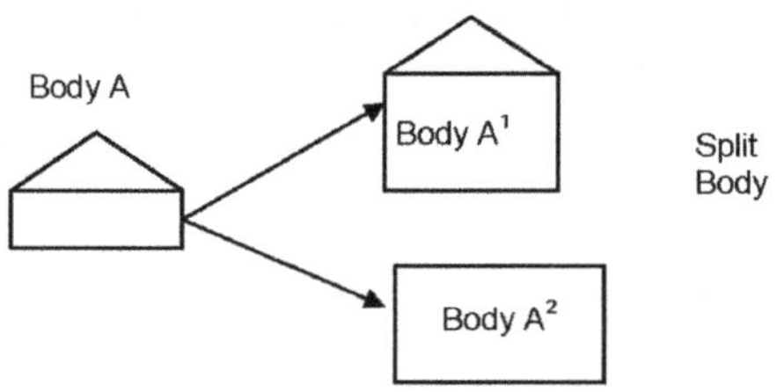

Or

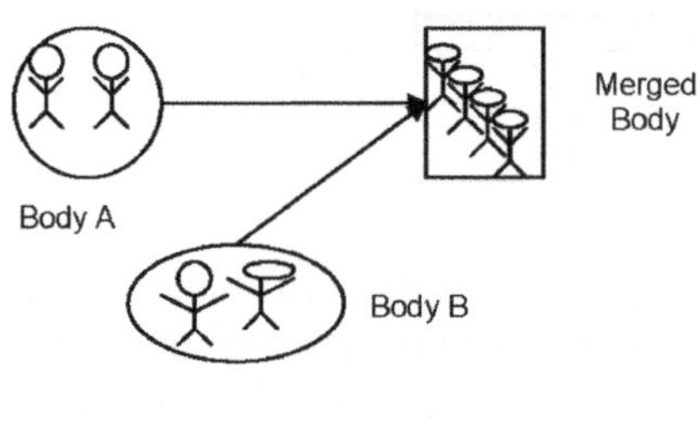

This relationship is usually expressed in authority data through see also references, information notes, and/or cataloguer's notes linking the authorized names given to each entity:

> The relationship between the corporate body known as Second Presbyterian Church, Cranbury, N.J. and the corporate body known as First Presbyterian Church, Cranbury, N.J., from which the former split in 1836 and later rejoined.
>
> The relationship between the musical groups known as Eppu Normaali, Popeda, and Räp Popeda which merged to form the musical group known as Eppupopedanormaali.
>
> The relationship between the corporate body known as Lapin korkeakoulu and the same corporate body later known as Lapin yliopisto.
>
> The relationship between the meeting known as Technical Thick Film Symposium and later meetings known as Symposium on Hybrid Microelectronics, Hybrid Microelectronics Symposium and International Hybrid Microelectronics Symposium.

5.3.7 Relationships among Works, Expressions, Manifestations, and Items [9]

Equivalence relationship

The subjective relationship between one manifestation or item and another that embodies the same intellectual or artistic content (work or work/expression), but are packaged or presented in different ways (e.g., an original and exact copies, digitizations, facsimiles, reprints, photocopies, microform versions, and similar other reproductions, such as an original printed map on paper and a photocopy or a microform copy or a digital copy of it, a sound recording on a disk and a CD of the same music, a motion picture of a reel and on a videocassette or DVD).

This is a type of shared characteristic, where the characteristic in common is the work. The equivalence is subjective because it depends on what the user deems to be the "same." However, this can be a very useful relationship to bring together all manifestations and items that embody the same work or work/expression.

This relationship is usually expressed in bibliographic data through the use of an authorized access point for the work in common (the shared characteristic), and/or through notes or through linked holdings data, rather than in authority data.

Derivative relationship

The relationship comprises two types. It includes relationships between an original work and a broad range of modifications from the original, including variations or versions of that work, such as editions, revisions, translations, summaries, abstracts, and digests. It also includes new works that are adaptations or modifications that become new works but are based on an earlier work (improvisations, etc.); new works that are changes of genre (transformations, dramatizations, novelizations, etc.); and, new works based on the style or thematic content of other works (free translations, paraphrases, imitations, parodies, etc.).

This relationship, also known as a "horizontal relationship", is usually expressed in authority data as the relationship between the authorized names given to each entity, and/or through information notes. It has also been expressed in bibliographic data through the use of an authorized access point for the original work with the extension of identifying elements to name the derived expression or manifestation.

Example

The relationship between the work known as Coriolanus by William Shakespeare and the work known as Coriolan von Shakespeare by Bertolt Brecht, which was adapted from the Shakespeare work.

The relationship between the work known as The Lord of the Rings by J.R.R. Tolkien and a parody of that work known as Bored of the Rings by Henry N. Beard.

The relationship between the work known as Ympäristöinformaatio kuluttajan päätöksenteossa and the work known as Environmental information in consumer decision making by Mari Niva, which is an English-language summary of the first work.

The relationship between the musical work known as La Chute de la Maison Usher by Claude Debussy and the literary work known as The Fall of the House of Usher by Edgar Allan Poe, on which the Debussy work was based.

The relationship between the work known as Roman d'Alexandre and the work known as Roman de toute chevalerie by Thomas of Kent, which is a later version of the first work.

Descriptive relationship

The relationship between a target work or expression or manifestation or item and a new work that refers to that target as its subject (e.g., criticism, commentary, review, casebook, etc., and the object of that analysis).

This type of relationship is usually expressed in bibliographic data through a subject controlled access point for the target work rather than through authority data.

Whole/part relationship

The relationship between a work and one or more works that form part of that larger work (e.g., the whole anthology and its individual works, the whole monographic series and its individual monographic parts, the whole serial and an individual special thematic issue or an article of importance).

This type of relationship, also known as a "vertical relationship" or "hierarchical relationship", is usually expressed in authority data through the use of see also references and/or information notes or cataloguer's notes. It also is often expressed through bibliographic data using the authorized access point for the work to identify the "whole" and/or notes to describe the relationship of the part to its whole; or through the use of contents notes and/or authorized access points for each of the parts to describe the relationship between the whole and its parts.

Example

The relationship between the work known as Anthologie grecque and the three parts of that work known as Anthologie palatine, Anthologie Planude, and Couronne.

Accompanying relationship (Part-to-Part)

The relationship between one work and another that is a **supplement** to that work (i.e., a work that is intended to be used in conjunction with or to augment the related work, such as a teacher's guide or incidental music to a play; or intended to **complement** or be a **companion** to another work, such as a set of maps to a video travelogue, a computer disk to accompany a textbook, a set of coloured plates to accompany a text, the equal parts of a kit).

This type of relationship is usually expressed in bibliographic data using an addition to the physical description, notes, and/or the authorized access point for related part, rather than through authority data. However, accompanying relationships can appear in authority data, especially when there are many manifestations involved, to provide see also references or information notes.

Examples

The relationship between the work known as the United States Catalog and the work known as the Cumulative Book Index, which is a supplement to the first work.

The relationship between the work known The Merchant of Venice by William Shakespeare and the work known as Incidental Music for Shakespeare's The Merchant of Venice by John Barbison.

Sequential relationship

The relationship between one work and another of different content that is a successor or a predecessor of that work (e.g., a sequel, a serial or series that has changed title). This excludes derivative works that modify the content of an earlier work (see Derivative relationship above).

This type of relationship, known as a “chronological relationship”, is usually expressed in bibliographic data through an authorized access point for the prequel and/or sequel and/or a note, or in authority data through a see also reference and/or an information note or cataloguer’s note. It could be indicated through the labelling of the authorized access point for the related works within an authority record for the “family of works.”

Example

The relationship between the work known as the Treaty of Amiens and an earlier treaty signed by Britain, France, Spain and the Batavian Republic, on which the former was based.

The relationship between the work Historical dictionaries of Africa and the same work later known as African historical dictionaries.

Shared characteristic relationship

The relationship between two works/expression/manifestations/items that have some characteristic in common, apart from any of the other relationships (e.g., same language, same color binding, same publisher, same date of issuance).

This type of relationship is expressed through the common access point or attribute in bibliographic or authority data for the entities involved.

5.4 Relationships between the Various Names of Persons, Families, Corporate Bodies, and Works

Within authority data, there are also a number of other relationship types that operate between specific instances of the entity *name* and other related names for the entities depicted in the upper part of **Figure 2** (*person*, *family*, *corporate body*, *work*, *expression*, *manifestation*, *item*, *concept*, *object*, *event*, and *place*).

Relationships in this category are normally reflected implicitly as name to name relationships by means of a “see” reference from a variant form to the authorized form of name or by means of "see also" references between related authorized forms of name for two or more related entities.

Table 2 highlights relationship types that operate between a specific instance of a name for a *person, family, corporate body,* or *work* and other names for that entity.

The relationship types identified in the table are intended to be representative only, and do not aim to provide an exhaustive taxonomy of relationships in this category that may be reflected in authority data. It is recognized that specific applications will select relationships important to the application, so there also is no intention to make these relationships mandatory or optional. (Note that relationships in this category specific to subject authority data are not included in the table.)

Definitions and examples for each of the relationship types identified in **Table 2** are given below in sections 5.4.1 – 5.4.4.

Table 2: Relationships between the Various Names of Persons, Families, Corporate Bodies, and Works

Entity Type	**Sample Relationship Types**
Name of Person ⇔ Name of Person	▪ earlier name relationship ▪ later name relationship ▪ alternative linguistic form relationship ▪ other variant name relationships
Name of Family ⇔ Name of Family	▪ alternative linguistic form relationship
Name of Corporate Body ⇔ Name of Corporate Body	▪ expanded name relationship ▪ acronym / initials / abbreviations relationship ▪ alternative linguistic form relationship ▪ other variant name relationships

Entity Type	Sample Relationship Types
Name of Work ⇔ Name of Work	▪ alternative linguistic form relation-ship ▪ conventional name relationship ▪ other variant name relationships

5.4.1 Relationships between Names of Persons

Earlier name relationship

The relationship between a name of a person and a name that person used at an earlier stage in life.

Examples

The relationship between a person who was made Earl of Beaconsfield, thus acquiring the name "Benjamin Disraeli, Earl of Beaconsfield" and that person's earlier name, "Benjamin Disraeli".

The relationship between a person who married a Mr. Nichols, thus acquiring the name "Charlotte Nichols" and that person's earlier name, "Charlotte Bronte".

Later name relationship

The relationship between a name of a person and a name that person used at a later stage in life.

Examples

The relationship between a name of a person who was first known as "Jacques-Anatole Thibeault" and that person's later name, "Anatole France".

The relationship between a person whose name prior to marrying Robert Browning was "Elizabeth Barrett" and that person's later name, "Elizabeth Barrett Browning".

Alternative linguistic form relationship

The relationship between a name of a person and an alternative linguistic form of name by which that person is known. This includes translations of a name into other languages.

Example

The relationship between the names of a person known as Horace in English and Quintus Horatius Flaccus in Latin.

Other variant name relationships

The relationship between a name of a person and several other variations on a name, such as orthographic relationships (spelling variations, transliterations, punctuation variations, capitalization variations), word order relationships (inversions, permutations), alternative name/synonym relationships (bynames, nicknames, courtesy names).

For some purposes it may be sufficient to simply identify all of these as "variant forms" of name for a person and to link them to the preferred form of name for the name of the person.

Examples

The relationship between the name of the person known as "Babe Ruth" and two other names by which that person is known, "Bambino" and "Sultan of Swat".

5.4.2 Relationships between Names of Families

Alternative linguistic form relationship

The relationship between a name of a family and an alternative linguistic form of name by which that family is known.

Example

The relationship between the names of the family known as Familie Trapp in German and Tappin perhe in Finnish.

5.4.3 Relationships between Names of Corporate Bodies

Expanded name relationship

The relationship between a name of a corporate body and the expanded form of name by which that body is known.

Example

The relationship between the name of the corporate body know as Abdib and the expanded form of that name, Associação Brasileira para o Desenvolvimento das Industrias de Base.

Acronym / initials / abbreviations relationship

The relationship between a name of a corporate body and an acronym or initials or an abbreviation by which that body is known.

Example

The relationship between the name of a corporate body known as the United Nations International Children's Emergency Fund and initialized form of that name, UNICEF.

Alternative linguistic form relationship

The relationship between a name of a corporate body and an alternative linguistic form of name by which that body is known. This includes translations of the name for the corporate body.

Example

The relationship between the names of the corporate body known as United Nations in English, Nations Unies in French and Организация Объединённых Наций in Russian.

Other variant name relationships

The relationship between a name of a corporate body and a name by which that body is known, such as orthographic relationships (spelling variations, transliterations, punctuation variations, capitalization variations), word order relationships (inversions, permutations).

Examples

The relationship between the name of the corporate body known as Women of Color Quilters Network and a spelling variant of that name, Women of Colour Quilters Network.

5.4.4 Relationships between Names of Works

Alternative linguistic form relationship

The relationship between a name of a work and an alternative linguistic form of name by which that work is known.

Example

The relationship between the names of the work known as Pentekostárion in transliterated Greek and Pentecostaire in French.

Conventional name relationship

The relationship between a name of a work and a conventional name by which that work is known.

Example

The relationship between the Beethoven symphony known as "Eroica" and the conventional title by which it is also known "Symphonies, no. 3, op. 55, E♭ major"

Other variant name relationships

The relationship between a name of a work and a name by which that work is known, such as orthographic relationships (spelling variations, transliterations, punctuation variations, capitalization variations), word order relationships (inversions, permutations), variant title relationships (titles for the work based on titles found on manifestations or items, including cover title, spine title, caption title, container title, parallel title).

Examples

The relationship between the names of the work known as Hamlet by William Shakespeare and the name that appears on some publications, Tragicall Historie of Hamlet Prince of Denmarke.

The relationship between the names of the work known as Honour This Day by Alexander Kent and the spelling variant of the name that appears on some publications, Honor This Day.

5.5 Relationships between Controlled Access Points

Within authority data there are also a number of other relationship types that operate between specific instances of the entity *controlled access point*. (i.e., between one access point and another, as distinct from relationships involving only the name or identifier that forms the basis for the controlled access point).

Relationships in this category may be made explicit by means of linking devices, such as linking fields and control subfields (e.g., a link between alternate script representations of the same controlled access point).

The relationships between controlled access points most commonly reflected in linking structure mechanisms are parallel language relationships, alternate script relationships, and different rules relationships.

Additional relationships exist between controlled access points for the names of the bibliographic entities and access points for the corresponding classification number or subject term for the same entity. Further links can also provide relationships to the digitized version of an entity through relating the name access point with the entity's identifier.

Table 3: Relationships between Controlled Access Points

Entity Type	Relationship Type
Controlled Access Point ⇔ Controlled Access Point	▪ Parallel language relationship ▪ Alternate script relationship ▪ Different rules relationship ▪ Name/corresponding subject term or classification number relationship ▪ Name/identifier relationship

Definitions and examples for each of those relationship types are given below.

Parallel language relationship

The relationship between two or more controlled access points that are established as parallel language forms of access point for the authorized forms of name for the same entity.

Example

The relationship between the controlled access point, Library and Archives Canada, as established in English, and the controlled access point, Bibliothèque et archives Canada, as established in French.

Alternate script relationship

The relationship between two or more controlled access points that are established as alternate script forms of the access point for the authorized forms of name for the same entity.

Example

The relationship between the controlled access point, Gogol, Nikolai Vasilievitch, expressed in roman alphabet, and the controlled access point, Гоголь, Николай Васильевич, expressed in Cyrillic alphabet.

Different rules relationship

The relationship between two or more controlled access points for the authorized forms of name for the same entity that have been established according to different sets of rules.

Example

The relationship between the controlled access point, Russia (Federation), established under one set of cataloguing rules, and the controlled access point, Российская Федерация, established under a different set of cataloguing rules.

Controlled access point/ Corresponding subject term or classification number relationship

The relationship between the controlled access point for the name of an entity (for a person, family, corporate body, work, expression, manifestation, item, concept, object, event, place) and a subject term in a controlled vocabulary, and/or a classification number for that entity. Here the subject term and the classification number may also be viewed as parallel names/identifiers for the name of the same entity.

Example

The relationship between the authorized access point, Library education, established according to Library of Congress Subject Headings practice, and the related Dewey Decimal Classification number, 020.7.

The relationship between the authorized access point, Gautama Buddha, as established according to AACR2, the authorized access point, Gautama Buddha, as established according to Library of Congress Subject Headings practice, and/or the related Library of Congress Classification number, BQ860-939.

Controlled access point/Identifier relationship

The relationship between the controlled access point for an entity expressed as an access point based on the authorized or variant form of name of the entity as related to an identifier for the entity.

Example

The relationship between the controlled access point, Advances in the history of rhetoric (Online), and the Linking ISSN for that work, 1536-2426.

6. User Tasks

For the purposes of this analysis, the users of authority data are broadly defined to include:

- authority data creators who create and maintain authority data;
- users who use authority information either through direct access to authority data or indirectly through the controlled access points (authorized forms of name, variant forms of name/references, etc.) in catalogues, national bibliographies, other similar databases, etc.

Four tasks representing all users are defined as follows:

Find	Find an entity or set of entities corresponding to stated criteria (i.e., to find either a single entity or a set of entities using an attribute or combination of attributes or a relationship of the entity as the search criteria); or to explore the universe of bibliographic entities using those attributes and relationships.
Identify	Identify an entity (i.e., to confirm that the entity represented corresponds to the entity sought, to distinguish between two or more entities with similar characteristics) or to validate the form of name to be used for a controlled access point.
Contextualize	Place a person, corporate body, work, etc., in context; clarify the relationship between two or more persons, corporate bodies, works, etc.; or clarify the relationship between a person, corporate body, etc., and a name by which that person, corporate body, etc., is known (e.g., name used in religion versus secular name).
Justify	Document the authority data creator's reason for choosing the name or form of name on which a controlled access point is based.

Table 4 maps the attributes and relationships defined in sections 4 and 5 to the defined set of user tasks. The mapping is intended to serve as a means of clarifying the specific user task(s) that data associated with each attribute or relationship defined in the model are designed to support. For the purposes of this conceptual model, all potential uses of data associated with a particular attribute or relationship are identified. There has been no attempt to assess or indicate the relative importance of each attribute or relationship to a given user task as was done in the FRBR model. Designations of which attributes and relationships are considered mandatory need to be determined on an application-specific basis.

Note: For the purposes of this study only attributes normally reflected in controlled access points are mapped against the FIND *task. The capability of conducting keyword searching on other data in the record is not taken into account. If it were, virtually any attribute could be mapped to* FIND.

Table 4: Mapping of Attributes and Relationships to User Tasks

Entity	**Attribute/Relationship**	**FIND**	**IDENTIFY**	**CONTEXTUALIZE**	**JUSTIFY**
Person	"Has appellation/is appellation of" relationship ⇔ name	■		■	
	"Assigned" relationship ⇔ identifier	■	■		
	Dates of person		■	■	■
	Title of person	■	■		

Entity	Attribute/Relationship	FIND	IDENTIFY	CONTEXTUALIZE	JUSTIFY
	Gender*		■		
	Place of birth*		■		
	Place of death*		■		
	Country*		■		■
	Place of residence*		■		■
	Affiliation*		■		
	Address*		■		
	Language of person*		■		■
	Field of activity*		■		
	Profession/occupation*		■		
	Biography/history*		■	■	
	Other information elements associated with the person	■	■		
	Pseudonymous relationship ⇔ person*	■		■	
	Secular relationship ⇔ person*	■		■	
	Religious relationship ⇔ person*	■		■	

Entity	Attribute/Relationship	FIND	IDENTIFY	CONTEXTUALIZE	JUSTIFY
	Official relationship ⇔ person*	■		■	
	Attributive relationship ⇔ person*	■		■	
	Collaborative relationship ⇔ person*			■	
	Sibling relationship ⇔ person*		■	■	
	Parent/child relationship ⇔ person*		■	■	
	Membership relationship ⇔ family*		■	■	
	Membership relationship ⇔ corporate body*			■	
	Pseudonym relationship ⇔ name of person*	■		■	
	Earlier name relationship ⇔ name of person*	■		■	
	Later name relationship ⇔ name of person*	■		■	
	Alternative linguistic form relationship ⇔ name of person*	■		■	
	Other variant name relationship ⇔ name of person*	■		■	

Entity	Attribute/Relationship	FIND	IDENTIFY	CONTEXTUALIZE	JUSTIFY
Family	"Has appellation/is appellation of" relationship ⇔ name*	■		■	
	"Assigned/is assigned to" relationship ⇔ identifier*	■	■		
	Type of family*	■	■		
	Dates of family*	■	■	■	
	Places associated with family*		■	■	
	Field of activity*		■		
	History of family*		■	■	
	Genealogical relationship ⇔ family			■	
	Founding relationship ⇔ corporate body*		■	■	
	Ownership relationship ⇔ corporate body*		■	■	
	Alternative linguistic form relationship ⇔ name of family*	■		■	

Entity	**Attribute/Relationship**	**FIND**	**IDENTIFY**	**CONTEXTUALIZE**	**JUSTIFY**
Corporate Body	"Has appellation/is appellation of" relationship ⇔ name	■		■	
	"Assigned/is assigned to" relationship ⇔ identifier	■	■		
	Place associated with the corporate body	■	■	■	
	Date associated with the corporate body	■	■	■	
	Language of the corporate body*		■		■
	Address*		■		
	Field of activity*		■		
	History*		■	■	
	Other information associated with the corporate body	■	■		
	Hierarchical relationship ⇔ corporate body*			■	
	Sequential relationship ⇔ corporate body*			■	
	Expanded name relationship ⇔ name of corporate body*	■		■	
	Acronym / initials / abbreviations relationship ⇨ name of corporate body*	■		■	

Entity	Attribute/Relationship	FIND	IDENTIFY	CONTEXTUALIZE	JUSTIFY
	Alternative linguistic form relationship ⇔ name of corporate body*	■		■	
	Other variant name relationship ⇔ name of corporate body*	■		■	
Work	"Has appellation/is appellation of" relationship ⇔ name	■		■	
	"Assigned/is assigned to" relationship ⇔ identifier	■	■		
	Form of work	■	■		
	Date of the work	■	■	■	■
	Medium of performance	■	■		
	Subject of the work*			■	
	Numeric designation	■	■		
	Key	■	■		
	Place of origin of the work*		■	■	■
	History		■	■	

Entity	**Attribute/Relationship**	**FIND**	**IDENTIFY**	**CONTEXTUALIZE**	**JUSTIFY**
	Other distinguishing characteristic	■	■		
	Equivalence relationship ⇔ work		■	■	
	Derivative relationship ⇔ work*			■	
	Descriptive relationship ⇔ work*			■	
	Whole/part relationship ⇔ work		■	■	
	Accompanying (part-to-part) relationship ⇨ work*			■	
	Sequential relationship ⇔ work*			■	
	Shared characteristic relationship ⇔ work*			■	
	Alternative linguistic form relationship ⇔ name of work*	■		■	
	Conventional name relationship ⇔ name of work*	■	■	■	
	Other variant name relationship ⇔ name of work*	■		■	

Entity	Attribute/Relationship	FIND	IDENTIFY	CONTEXTUALIZE	JUSTIFY
Expression	"Has appellation/is appellation of" relationship ⇔ name	■	■		
	"Assigned/is assigned to" relationship ⇔ identifier	■	■		
	Form of expression	■	■		
	Date of expression	■	■		
	Medium of performance of expression	■	■		
	Language of expression	■	■		
	Technique*		■	■	
	Other distinguishing characteristic	■	■		
Manifestation	"Has appellation/is appellation of" relationship ⇔ name	■	■		
	"Assigned/is assigned to" relationship ⇔ identifier	■	■		
	Edition/issue designation		■		
	Place of publication/ distribution		■		
	Publisher/distributor		■		

Entity	**Attribute/Relationship**	**FIND**	**IDENTIFY**	**CONTEXTUALIZE**	**JUSTIFY**
	Date of publication/distribution		■		
	Form of carrier		■		
	Numbering		■		
Item	"Has appellation/is appellation of" relationship ⇨ name	■	■		
	"Assigned/is assigned to" relationship ⇨ identifier	■	■		
	Location of item*		■		
	Custodial history of item		■		
	Immediate source of acquisition of item		■	■	
Concept	"Has appellation/is appellation of" relationship ⇔ name	■		■	
	"Assigned/is assigned to" relationship ⇔ identifier	■	■		

Entity	Attribute/Relationship	FIND	IDENTIFY	CONTEXTUALIZE	JUSTIFY
Object	"Has appellation/is appellation of" relationship ⇔ name	■		■	
	"Assigned/is assigned to" relationship ⇔ identifier	■	■		
Event	"Has appellation/is appellation of" relationship ⇔ name	■		■	
	"Assigned/is assigned to" relationship ⇔ identifier	■	■		
Place	"Has appellation/is appellation of" relationship ⇔ name	■		■	
	"Assigned/is assigned to" relationship ⇔ identifier	■	■		
Name	Type of name	■	■	■	
	Name string	■	■	■	
	Scope of usage			■	

Entity	Attribute/Relationship	FIND	IDENTIFY	CONTEXTUALIZE	JUSTIFY
	Dates of usage			■	
	Language of name		■		
	Script of name		■		
	Transliteration scheme of name		■		
	"Is basis for/is based on" relationship ⇔ controlled access point	■			
Identifier	Type of identifier		■		
	"Is basis for/is based on" relationship ⇔ controlled access point	■			
Controlled Access Point	Type of controlled access point		■		
	Status of controlled access point for the preferred form		■		■
	Designated usage of controlled access point		■	■	
	Undifferentiated access point		■		

Entity	Attribute/Relationship	FIND	IDENTIFY	CONTEXTUALIZE	JUSTIFY
	Language of cataloguing		■		■
	Script of base access point		■		
	Script of cataloguing		■		■
	Transliteration scheme of base access point		■		
	Transliteration scheme of cataloguing		■		■
	Source of controlled access point		■		■
	Base access point	■	■		
	Addition	■	■	■	
	"Is governed by/governs" relationship ⇔ rules		■		■
	"Is created/modified by/creates/modifies" relationship ⇔ agency		■		■
	Parallel language relationship ⇔ controlled access point for the authorized form			■	
	Alternate script relationship ⇔ controlled access point for the authorized form			■	
	Different rules relationship ⇔ controlled access point for the authorized form			■	

Entity	Attribute/Relationship	FIND	IDENTIFY	CONTEXTUALIZE	JUSTIFY
	Controlled access point/Corresponding subject term or classification number relationship ⇔ controlled access point	■	■	■	
	Controlled access point/Identifier relationship ⇔ controlled access point	■	■		
Rules	Citation for rules		■	■	■
	Rules identifier		■		
	"Are applied by/applies" relationship ⇔ agency				■
Agency	Name of agency		■		
	Agency identifier		■		
	Location of agency		■		

References

1. Working Group – Functional Requirements for Subject Authority Records (FRSAR). www.ifla.org/en/frsar-wg (accessed May 11, 2009)

2. Guidelines for authority records and references / revised by the Working Group on GARE Revision. Second edition. München : K.G. Saur, 2001.

3. Functional requirements for bibliographic records : final report / IFLA Study group on the Functional Requirements for Bibliographic Records. München : K.G. Saur, 1998.

4. Statement of international cataloguing principles. 2009. www.ifla.org/files/cataloguing/icp/ICP-2009_en.pdf (accessed May 11, 2009)

5. UNIMARC manual – authorities format. 2nd revised and enlarged edition. München : K.G. Saur, 2001. www.ifla.org/files/uca/unimarc-authorities-format.pdf (accessed May 11, 2009) (Note: This on-line version includes corrections made in 2004.)

6. Mandatory data elements for internationally shared resource authority records : report of the IFLA UBCIM Working Group on Minimal Level Authority Records and ISADN. Frankfurt am Main, 1998. http://archive.ifla.org/VI/3/p1996-2/ mlar.htm (accessed May 11, 2009)

7. International standard archival authority record for corporate bodies, persons and families. 2nd edition. April 2004. www.icacds.org.uk/eng/isaar2ndedn-e_3_1.pdf (accessed May 11 2009)

8. Encoded archival context. November 2004. www.iath.virginia.edu/eac/ (accessed May 11, 2009)

9. Based on Tillett, Barbara B. "Bibliographic Relationships," in C.A. Bean and R. Green (eds.), *Relationships in the Organization of Knowledge*. Dordrecht : Kluwer Academic Publishers, 2001, p. 19-35.

Index

This index covers only terms used for the entities, attributes and relationships defined in this model. The page number following the term indicates the page on which the term is defined.

Printed and bound by CPI Group (UK) Ltd, Croydon, CR0 4YY
13/09/2026
14960852-0006